Birds of the West Virginia Northern Panhandle

Philip M. Carter

75 years ago, George Miksch Sutton wrote an article on the birds of the West Virginian Northern Panhandle (Sutton, 1933). Forty-two years later, Dr. A.R. Buckelew updated Mr. Sutton's list. The goal of the author of this field guide companion is to not only update the information that came before me but to expand on it. Dr. Buckelew's original guide, "The Red Start" publication, the Cornell Lab of Ornithology's "E-bird" database and "The West Virginia Breeding Bird Atlas" also by Dr. Buckelew and George Hall will be used as reference for their historical records to show how the biological diversity of the Northern Panhandles' avifauna has changed over the last 113 years.

Agriculture which was once a thriving industry in the Panhandle has severely declined and has been replaced by industrial enterprise which gave way to consumptive natural resource use. Coal mining, natural gas drilling or "fracking" as it is also known are now the large industries of the Panhandle along with logging, and other natural resource dependent business. The result of both natural resource reliance and a large number of new commercial buildings coming into the Panhandle is habitat loss. Beech Bottom Swamp which was once home to many species of interest including many that were only found there in the Panhandle is no longer there. However, not all habitats have been lost, and some birds which have been absent over the years have been able to move back into the Panhandle. Species that favor woodlands such as the Red-headed Woodpecker have been on the incline. Though species such as Lark Sparrows, Dickcissels, and Bewick's Wrens are now just as rare as when Dr. Albert Buckelew wrote his guide in 1976. As with the previous edition of this guide, many new species ranging from accidental to casual have been added to the list panhandle list. Rufous Hummingbirds for example which were absent from the state are now uncommon migrants.

GEOGRAPHY

The West Virginia Northern Panhandle consists of four counties. Marshall, located in the south. Ohio, and Brooke, located in the center. Hancock, located in the north. The Eastern, and Northern boundaries of the state are marked by the Ohio River. The Ohio river is part of West Virginia until the Ohio state shoreline. The Eastern boundary is the Pennsylvania state line. The main body of the state of West Virginia is located south of Marshall county. Major towns include: New Cumberland in Hancock County, Wellsburg in Brooke County, Wheeling in Ohio County, and Moundsville in Marshall County. The length of the Northern Panhandle is 60 miles long, and the width ranges from five to eighteen miles across. The climate is humid and continental and is characterized by a seasonal difference in temperature. The average daily temperature is 63° F, and the average daily minimum temperature is 42° F. The Average annual minimum temperature is -3° F, and the average annual maximum temperature is 95° F. The Annual precipitation is evenly distributed with an average of 18 to 24 inches per year. Average snowfall per year can vary from twelve inches a year to around 24 inches per year which is the average amount for the Northern Panhandle. There are several lakes all of which are manmade in the Northern Panhandle: Bear Rock Lakes in Ohio County, Castleman's Run Lake in Brooke, Schenk Lake in Ohio County, and Tomlinson Run Lake in Hancock County. Along with the previously noted lake numerous ponds. Several small lakes can be found along the Ohio River which are made by the navigational locks in Hancock County. Swampland is scarce due to the number of hills that allow water to be drained efficiently. Small swamps can be found at inlets however, such as at Castleman's Run Lake, and Tomlinson Run Lake. A majority of the hills in the Panhandle are around 1100 feet in altitude while some can in fact range as high as 1600 feet in elevation.

ANNOTATIONS
Frequency of individuals to be expected in the appropriate habitat and in each season are indicated as follows:
Accidental - Fewer than 4 or 5 Records
Casual - Species which cannot be expected every year or even in a period of several years.
Rare - One to Ten Can be expected most years
Uncommon - One to five can be seen daily
Fairly Common - Five to Ten Per Day
Common - 10 to 50 Per Day
Very Common - 50 to 200 Per Day
Abundant - More than 200 Per Day.

Seasonal Occurrences

Permanent Resident - A species which is found throughout the year
Summer Resident - A species which breeds in the Panhandle but migrates out of the area in winter.
Migrant - A species which is observed in the migration season.
Winter Visitor - A species which nests elsewhere, but is found during the winder in the Panhandle.
Extinct and exotic species are not included.

CLASSIFICATION AND NOMENCLATURE

The American Ornithologists Union (A.O.U) Checklist of North and Middle American Birds (2014) and the Fifty-fourth Supplement to the A.O.U checklist (2013) are the standards that were used for the classification and nomenclature for this guide.

FINDING GUIDE

It is this authors hope that this field guide companion will help students who are new to Ornithology find new species, and to help locate known species for study, as well as enthusiasts with experience to have the chance to add to their lists. Whenever possible, I have included the location of reports such as parks. The general habitat or specific location for a species, as well as proper season are also provided when possible.

WILDLIFE MANGAMEMENT AREAS

Thanks to the help of the West Virginia Department of Natural Resources I have been given permission to use maps of the Wildlife management areas located in the Northern Panhandle. WMA's located in the Northern Panhandle: **Hancock County**: Hillcrest WMA. **Brooke County**: Cross Creek WMA, Castlemans Run Lake WMA located in both Brooke and Ohio County. **Ohio County**: Bear Rock's lake WMA. **Marshall County**: Burches Run WMA, Cecil H. Underwood WMA, and Dunkard Fork WMA. All of these locations provide excellent birding opportunities and I highly recommend to the reader that you explore these areas. For in depth accounts and guides for each WMA, the Brooks Bird Club has published a guide entitled "Birding Guide to West Virginia" of which the 2nd edition has recently been released and was compiled by Greg E. Eddy. For more information about this guide or the Brooks Club, the club can be contacted online at www.brooksbirdclub.org

or by mail at The Brooks Bird Club of West Virginia, P.O. Box 4077, Wheeling, WV 26003.

ACKNOWLEDGEMENTS

First and foremost, I want to thank Dr. Albert Buckelew who has been a mentor and an outstanding friend for giving me this opportunity. It is an honor to follow in both he, and George Suttons footsteps by writing this guide. I also want to thank the Brooks Bird Club and its members especially who read early drafts of this book and provided amazing feedback. I would also like to thank my parents Philip R. Carter, and Kimberly A. Carter as well as former Pennsylvania Fish Commission employee Richard Ketchum, and Steve Irwin the Crocodile Hunter for instilling in me from an early age a love of both the outdoors and nature. I would especially like to thank the West Virginia Department of Natural Resources for providing and allowing me to use the maps located in this book. Along with the previously mentioned individuals and groups I would also like to thank the Cornell Lab of Ornithology for graciously allowing me access to historical records from the last 100 years in the West Virginia Northern Panhandle. Without their database this project would nave have been possible. Last but not least, I want to thank you the reader, without you and others like yourself using this guide, my job would only not be harder, but it would be almost impossible.

COUNTY MAPS

WMA KEY
1 – Bear Rock Lakes WMA
3 – Burches Run WMA
4 – Castleman's Run Lake WMA
5 – Cecil H. Underwood WMA
7 – Cross Creek WMA
8 – Dunkard Fork WMA
9 - Hillcrest WMA.

WILDLIFE MANAGEMENT AREA MAPS

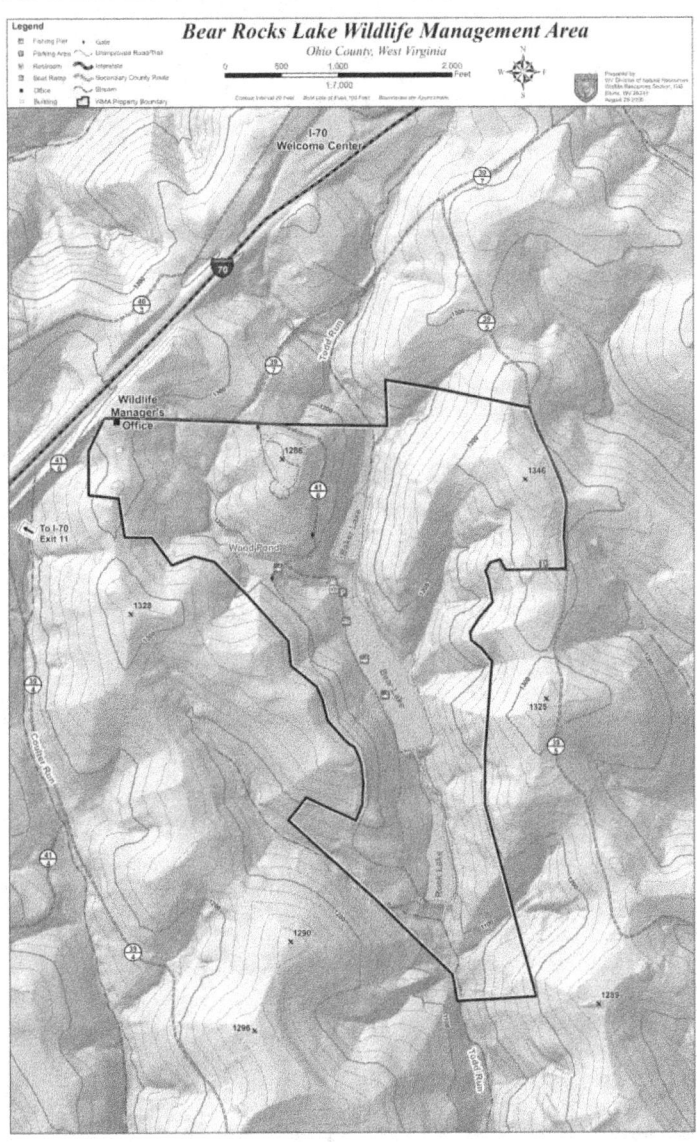

Castlemans Run Lake Wildlife Management Area

Brooke & Ohio counties, West Virginia

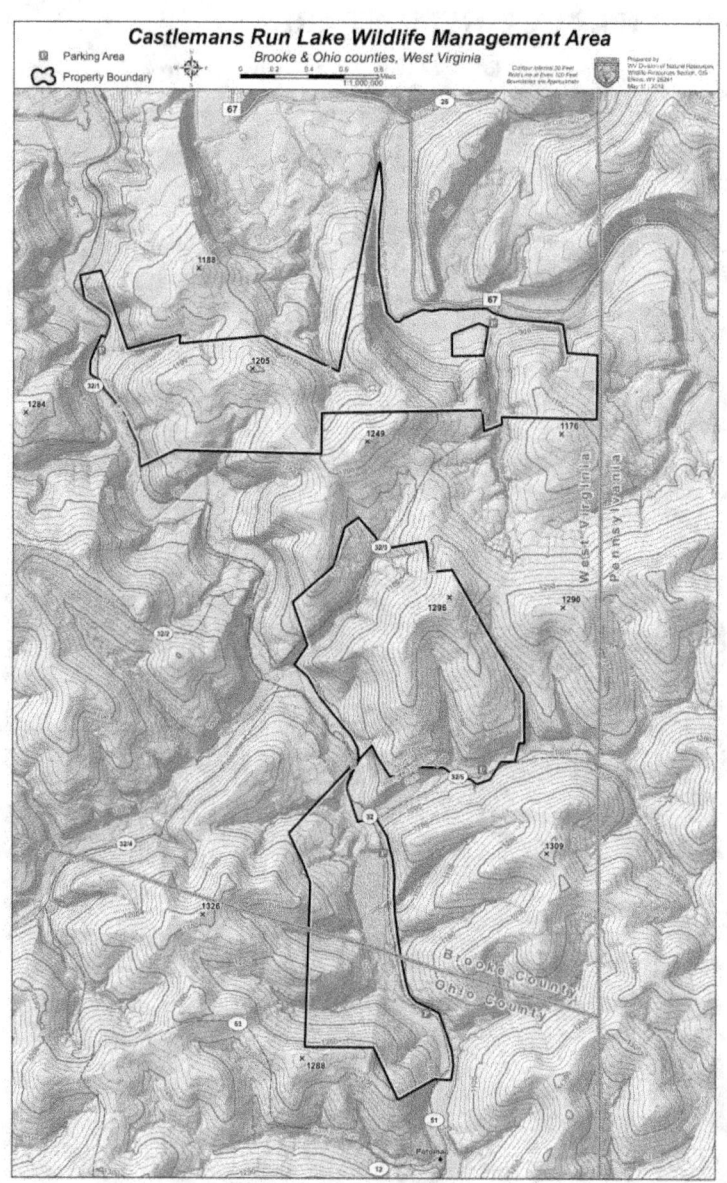

Parking Area

Property Boundary

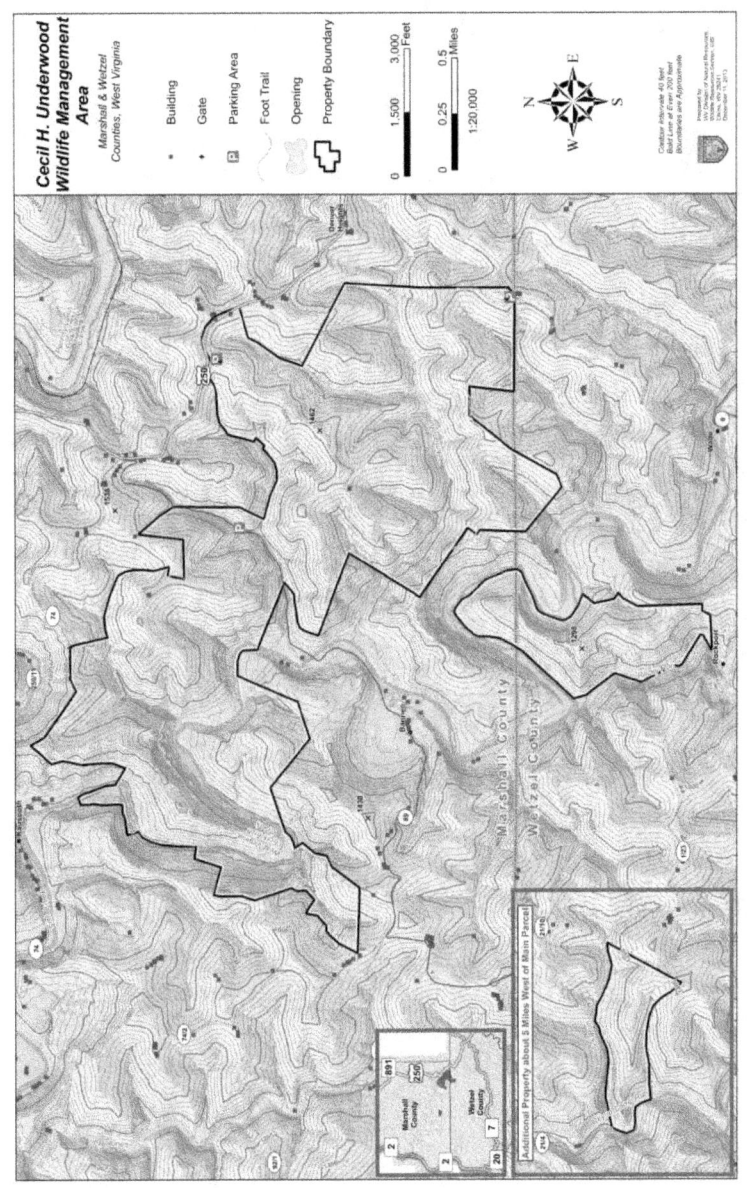

Cecil H. Underwood Wildlife Management Area

Marshall & Wetzel Counties, West Virginia

- • Building
- • Gate
- ▣ Parking Area
- Foot Trail
- Opening
- Property Boundary

0   1,500   3,000 Feet

0   0.25   0.5 Miles

1:20,000

N E S W

Contour interval 40 feet
Bold Line at Even 200 feet
Boundaries are Approximate

Prepared by:
WV Design of Natural Resources
Wildlife Resources Section GIS
Elkins, WV 26241
December 1, 2013

Additional Property about 5 Miles West of Main Parcel

Marshall County

Wetzel County

Cross Creek Wildlife Management Area
Brooke County, West Virginia

	Parking Area
	Gate
	Wildlife Clearing
	WMA Property Boundary

Contour Interval 20 Feet
Bold Line at Even 100 Feet
Boundaries are Approximate

0   0.25   0.5   0.75   1
Miles

1:18,000

Prepared by:
WV Division of Natural Resources,
Wildlife Resources Section, GIS
Elkins, WV 26241
November 27, 2012

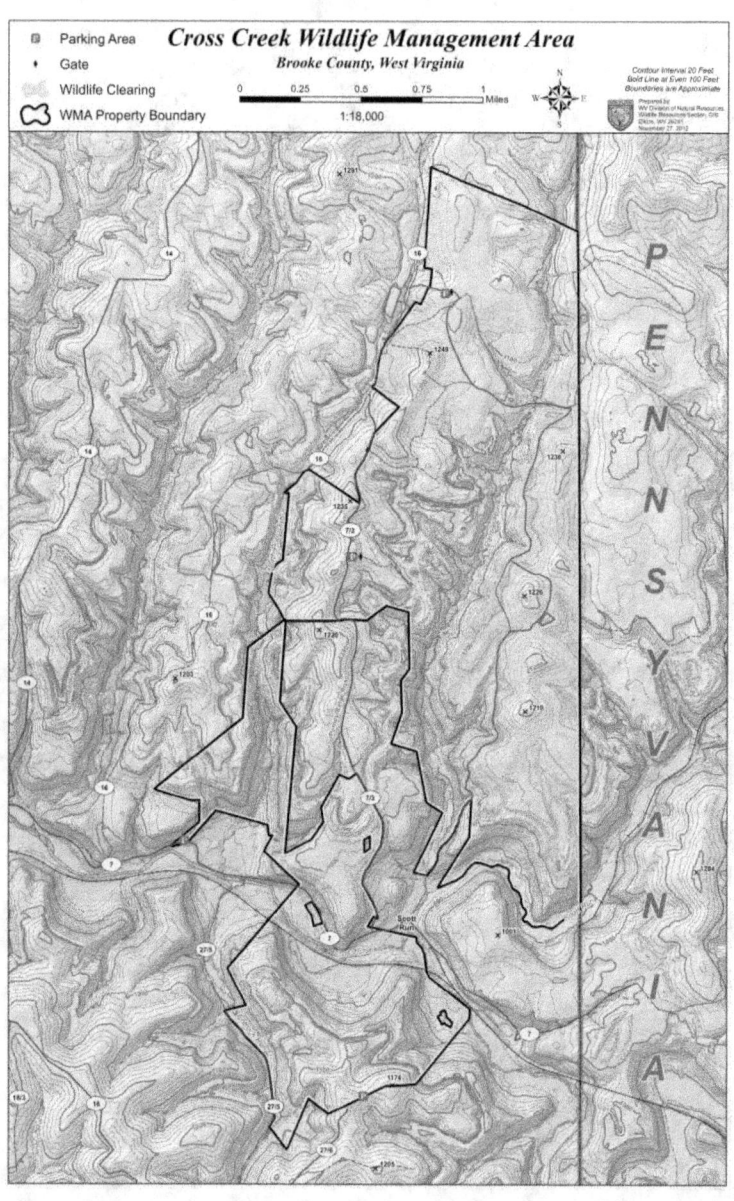

Dunkard Fork Wildlife Management Area
Marshall County, West Virginia

0   750   1,500   2,250   3,000 Feet

1:10,000

Contour Interval 40 Feet
Bold Line at Even 200 Feet
Boundaries are Approximate

Prepared by
WV Division of Natural Resources
Wildlife Resources Service, WB
Elkins, WV 26241
November 27, 2012

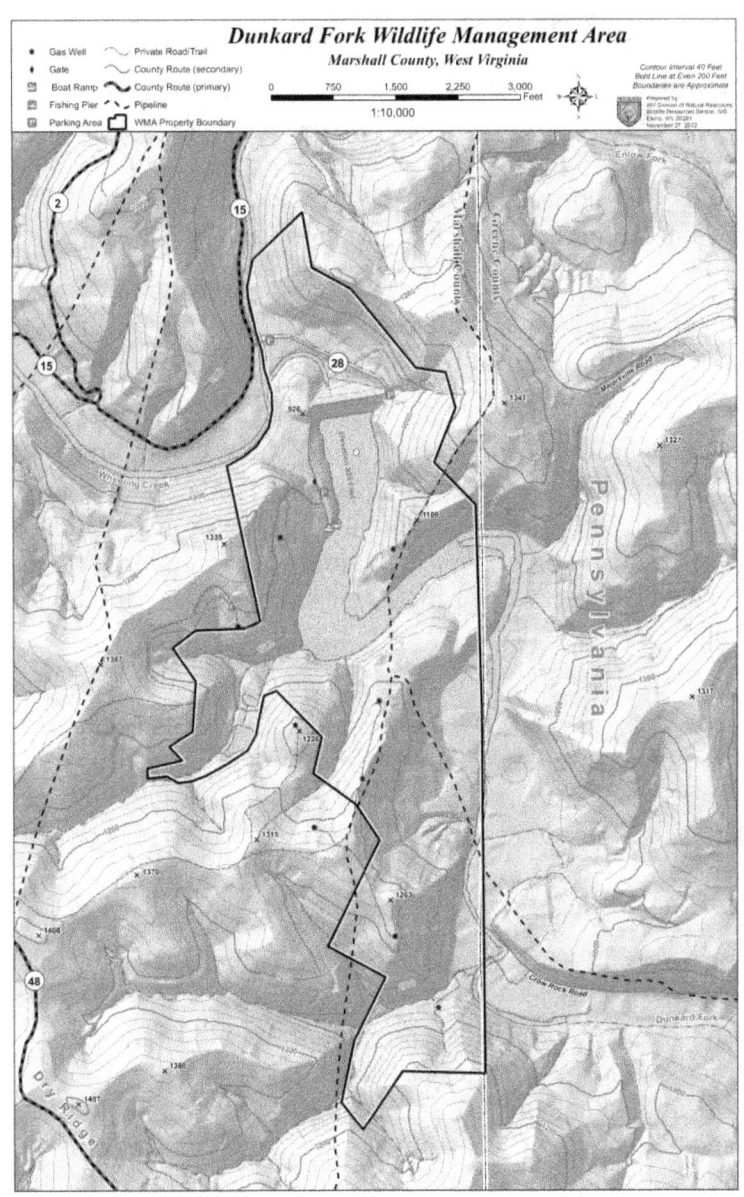

Checklist of WV Birds

Anatidae

Greater White-fronted Goose *(Anser albifrons)* – Accidental. Outside of Panhandle, casual. Could be more common, due to lack of reports. John Boback reported one February 27th, 2013 at the Pike Island Dam, in Ohio County.

Snow Goose *(Chen caerulescens)* – Casual Migrant. Sometimes seen on Bear Rock Lakes, Castleman's Run Lake and Tomlinson Run Lake. The Blue Goose, a dark morph of the Snow Goose is seen less frequently.

Canada Goose *(Branta canadensis)* – Common migrant to Permanent Resident. Depending on location Canadian Geese can be seen all year round. Bethany College is home to a permanent breeding population. Large migrating flocks can also be observed during some years passing through the Panhandle.

Mute Swan *(Cygnus olor)* – Accidental. A feral specimen was taken at Wellsburg, on December 13, 1934 (Buckelew, 1976). Recent reports of this species along the Ohio River near Wheeling-Ohio Airport indicate this could be a casual migrant.

Tundra Swan *(Cygnus columbianus)* – Accidental migrant. As with the previous accidental species, outside of the Panhandle this is a casual migrant, and could be the same inside the panhandle due to lack of reports. Two records exist in the Panhandle of this bird. Both are from Ogelbay Park. The first report is from Scott Albaugh on December, 5, 2012 where he reports a flock of 150. The second report comes from Wilma Jarrell on December 22, 2012 where a flock of 30 were reported.

Wood Duck *(Aix sponsa)* – Fairly common migrant and uncommon summer resident. Breeding pairs can be found along all the major streams and on lakes in remote areas. This species as of the West Virginia Breeding Bird Atlas II has been confirmed to breed in nearly all counties in the Northern Panhandle. I observed a probable breeding pair in 2013 along the Buffalo Creek near Bethany College.

Gadwall *(Anas strepera)* – Accidental/Casual migrant. Can be found along the Ohio River, lakes and larger streams of the Panhandle. Reports have been less common in recent years with less than 5 reported to E-bird.

American Wigeon *(Anas americana)* – Accidental/ Casual migrant. Almost exclusively found along the Ohio River, as with the Gadwall, reports to E-bird have become extremely scarce of this species.

American Black Duck *(Anas rubripes)* – Casual migrant. Possibly a rare migrant due to accounts in recent years. Formerly a fairly common migrant, records since 1976, have all but stopped until around 2009. Historically this species was found along the Ohio River, and lakes inland. It has also bred in Hancock County where. E.R. Chandler observed it throughout the summer on Tomlinson Run fishing ponds. (Buckelew, 1976). However, no reports of this species breeding in Northern Panhandle exist as of the second Breeding Bird Atlas.

Mallard (*Anas platyrhynchos*) – Fairly common migrant and permanent resident. Seen along larger streams, ponds, lakes, and the Ohio River. Big Wheeling Creek at Elm Grove has historically been an excellent location to observe large flocks (Buckelew, 1976). This species breeds extensively in the Northern Panhandle.

Blue-winged Teal (*Anas discors*) – Casual migrant. Rare summer resident. Less than 5 records though E-bird currently exist. During the spring 2013 semester, I observed one during Ornithology class near Bethany College football field. Species has nested at Bear Rocks (Buckelew 1976).

Northern Shoveler *(Anas clypeata)* – Accidental, possible Casual migrant. Has been found along the Ohio River as well as larger stream and lakes. Reports have become scarce.

Northern Pintail *(Anas acuta)* – Accidental/Casual migrant. Found along the Ohio River, less than 5 reports are recorded to E-bird.

Green-winged Teal *(Anas crecca carolinensis)*- Accidental/Casual Migrant. Three records are available through E-bird. The first being from West Virginia historical records in 1963, with the latest two reports taking place on March 3, 2012, and February 27, 2013. Reports outside the Panhandle indicate this species could be a casual migrant due to absence of reports.

Canvasback (*Aythya valisineria*) – Casual winter migrant. When present this species is can be found along the Ohio River.

Redhead (*Aythya americana*) – Casual winter migrant. When present this species is can be found along the Ohio River. Flocks on the Ohio side of the Ohio River near East Liverpool, OH have been reported of up to 100 ducks.

Ring-necked Duck *(Aythya collaris)* – Rare winter migrant. Found along the Ohio River, I witnessed two near Bethany on June 22, 2013. Possibly a migrating pair or a pair that wintered. One was in fact reported to have wintered on Schenk Lake, in Oglebay Park in 1973-74 (Buckelew, 1976).

Greater Scaup (Aythya marila) – Casual Migrant along the Ohio River.

Lesser Scaup *(Aythya affinis)* – Casual Migrant along the Ohio River.

White-winged Scoter *(Melanitta fusca)* – Accidental Migrant along the Ohio River.

Long-tailed Duck (Oldsquaw) *(Clangula hyemalis)*- Accidental Migrant along the Ohio River.

Bufflehead *(Bucephala albeola)* – Rare migrant along the Ohio River.

Common Goldeneye *(Bucephala clangula)* – Casual winter migrant along the Ohio River. I observed one along the Buffalo Creek near Bethany on April 16. 2013.

Hooded Merganser *(Lophodytes cucullatus)* – Rare winter migrant along the Ohio River, rare spring migrant. Most reports are along the Ohio River but I observed a flock of 8 Hooded Mergansers on the Buffalo Creek near Bethany on

May 8, 2013.

Common Merganser *(Mergus merganser)* – Casual/Rare winter migrant found along the Ohio River and larger streams.

Red-breasted Merganser *(Mergus serrator)* – Rare winter migrant. Found along the Ohio River.

Ruddy Duck *(Oxyura jamaicensis)* – Accidental/ Rare winter migrant found along the Ohio River.

Odontophoridae
Northern Bobwhite (*Colinus virginianus*) – Accidental possible permanent resident. Two historical records in 1960, and 1975 exist both near Wheeling. Another more recent account from Sandi Wheeler on June 2, 2013 is from Hancock county in Chester, WV. These birds are extremely rare in the Panhandle and surrounding areas.

Phasianidae
Ring-necked Pheasant (*Phasianus colchicus*) – Rare resident. Found mainly in agricultural areas where fields are present.

Ruffed Grouse *(Bonasa umbellus)* – Rare resident. Subject to cyclical fluctuations in population. Look for slopes with a heavy underbrush to improve chances of observation (Buckelew, 1976).

Wild Turkey (*Meleagris gallopavo*) – Uncommon/ Fairly common permanent resident depending on location. Population has made a rebound since the 1976 publication.

Most likely as Dr. Buckelew suspected, this was from stocked Turkey's both here in West Virginia as well as Pennsylvania.

Gaviidae
Red-throated Loon (*Gavia stellata*) – Accidental. Only one record exists in the E-bird database. On June 1, 1966 the Temple family observed one on the Ohio River near Wheeling, WV.

Common Loon (*Gavia immer*) – Casual/Rare winter migrant along the Ohio River and Casual inland. The specimen on display at Bethany College was taken along Castleman's Run (Sutton, 1939)

Podicipedidae
Pied-billed Grebe *(Podilymbus podiceps)* – Rare migrant. Found along the Ohio River, and occasionally at Ogelbay Park, as well as lakes inland.

Horned Grebe *(Podiceps auritus)*- Casual/Rare migrant and winter visitor. Most often found along the Ohio River, and can occasionally be observed at Castleman Run Lake near Bethany, WV.

Red-necked Grebe (*Podiceps grisegena*) – Accidental. One found frozen by breast feathers to the ice of the Buffalo Creek on Feb. 17, 1934. William Montagna identified and then mounted it. (Buckelew, 1976).

Eared Grebe *(podiceps nigricollis)* – Accidental. Tom Shields reported one at Wheeling Island on February 12, 1955 (Buckelew, 1976).

Phalacrocoracidae
Double-crested Cormorant *(Phalacrocorax auritus)*- Casual winter migrant. Dead specimen found at Warwood, Oct. 22, 1951 (Buckelew, 1976). While other sporadic reports do exist, very large gaps sometimes of decades do exist for this bird.

Ardeidae
American Bittern *(Botaurus lentiginosus)* – Casual migrant. Historical records indicate this species may have nested at Beech Bottom Swamp (Sutton, 1933). Charles Conrad saw 9 in the air once over Beech Bottom Swamp in September 1934 (Buckelew, 1976). In addition to these records, two other records exist, one in 1957 from Wellsburg, and another in 1970 from Chester, in Hancock County.

Least Bittern *(Ixobrychus exilis)* – Casual migrant. May have nested at one time at Beech Bottom Swamp (Sutton,1933). Swamps in areas such as Castleman's Run Lake, Tomlinson Run Park and others should be checked very carefully for both this species and the American Bittern (Buckelew,1976)

Great Blue Heron *(Ardea herodias)* – Fairly common summer visitor. Some individuals do stay into early winter as indicated by records from the Hancock County Christmas Count (Laitsch, 1971). A large colony can be found on the north end of Brown's island. Best observation location is from the Ohio side of the river in Toronto, Ohio (Buckelew, 1976). Along with these locations most lakes, ponds, and streams are good areas to find Great Blue Herons as well.

Great Egret *(Ardea alba)* – Casual summer migrant. Only

four records for the Pandhandle exist through E-bird with less than five individual birds reported. The last record is from October of 2011 in which Thomas Czubek observed at least one near Middle Grave Creek in Marshall County two days in a row.

Snowy Egret *(Egretta thula)* – Accidental. Two records exist for the bird in the Panhandle. One from 1956 which was reported by T.Shields (Redstart 23,4). The other is a record from May, 1975 which was reported by the Temple family (Buckelew, 1976).

Little Blue Heron *(Egretta caerulea)* - Casual summer visitor. Could possibly be accidental as only four records in the Panhandle exist. Three out of four records are from the 1930's with the last report being from 1960. Two reports are from Wheeling, with the other two are from Triadelpha along the Big and Little Wheeling Creek (E-bird, 2013).

Cattle Egret *(Bubulcus ibis)*- Accidental. Nevada Laitsch reported one on May 12-14, 1974 in Hancock County (Buckelew, 1976).

Green Heron *(Butorides virescens)* – Rare summer resident. Nests along small streams. Bethany College has had one periodically in the pond by the football field. Castlman's Run Lake and ponds location along Rt.2 in Hancock are also good areas for observation.

Black-crowned Night-Heron *(Nycticorax nycticorax)* - Casual migrant. Could possibly be accidental. Very few records exist, almost all of which are along the Ohio River in the Wheeling area.

Yellow-crowned Night-Heron *(Nyctanassa violacea)* –
Accidental. A.B. Brooks examined a specimen which had
been taken near Wheeling in 1944 (Brooks, 1944). One
other record according to E-bird exists from 1981 and was a
mid-month estimate for date: Observed several times during
that month by Dr. Buckelew.

Cathartidae
Turkey Vulture *(Cathartes aura)* – Common permanent
resident. Nests in the Panhandle. Can be found along open
areas such as field, and road ways when carrion is present.
Can roost in large numbers.

Pandionidae
Osprey *(Pandion haliaetus)* – Casual winter migrant.
Historical data shows that this species has nested at both
Tomlinson Run State Park and Bear Rocks. The highest
number or records are again from the Wheeling area along
the Ohio River.

Accipitridae
Bald Eagle *(Haliaeetus leucocephalus)* – Casual migrant
along the Ohio River. Reports are becoming more frequent
as nests are now being found. A pair has been nesting along
the Ohio River at the Pike Island Dam. Other nests
throughout the Panhandle have also been reported. Juveniles
have also been observed. I, myself observed a juvenile
flying over Bethany Community Park on July 29, 2013 not
far from where a photo of an adult was taken a few months
prior.

Northern Harrier *(Circus cyaneus)* – Casual/Rare migrant.
Low wet fields such as those by Bethany along Buffalo
Creek are good areas to look. I witnessed one being mobbed

by crows by the B&G building of Bethany College, not far from the Buffalo Creek.

Sharp-shinned Hawk (*Accipiter striatus*) – Uncommon migrant and winter visitor. Rare summer resident. Along with the Cooper's Hawk this species of *Buteo* will take birds from feeders. Historical records show nests have been found near Ogelbay and Bethany however, not since the 1930's (Sutton, 1933).

Cooper's Hawk (*Accipiter cooperi*) – Uncommon permanent resident. Nests at both Ogelbay and Parkinson Forest at Bethany College.

Northern Goshawk (*Accipiter gentilis*) – Accidental /Casual winter visitor. Only three records exist, Christmas Eve from 1919 near Follansbee reported by George Miksch Sutton. Nov. 18, 1933 from Oglebay Park by R. Rine, and T.Shields, finally, December 12, 1974 Dr. A.R. Buckelew Jr. reported an individual in Bethany.

Red-shouldered Hawk *(Buteo lineatus)* – Casual migrant and winter visitor.

Broad-winged Hawk (*Buteo platypterus*) – Rare migrant and summer resident. More common in both the Northern and Southern parts of the Panhandle. In the Northern section of the Panhandle often reported as the Red-tailed Hawk.

Red-tailed Hawk (*Buteo jamaicensis*) – Fairly common permanent resident. Reports since the 1970's now indicate this species is present all year round as to being more commonly present in the summer. Nests are large and obvious when found, they have been found in nearly all parts of the Panhandle. Easily mistaken for Broad-winged

Hawk.

Rough-legged Hawk (*Buteo lagopus*) – Casual winter visitor. Observed by several members of the George Sutton Audubon Society during December of 1974 near the town of Bethany (Buckelew, 1976).

Rallidae
Yellow Rail *(Coturnicops noveboracensis)* – Accidental. One record from Ohio County (Brook, 1944)

King Rail *(Rallus elegans)* – Accidental/Casual migrant. May not be as common as once was. No data was found from 1977 through 2013 from Cornell's E-bird database, species may no longer visit the Northern Panhandle.

Virginia Rail (*Rallus limicola*) – Casual/Rare migrant. Charles Conrad located a nest at the now non-existant Beech Bottom Swamp in 1936 (Buckelew, 1976). Several reports exist from 1933, and one from May 27, 1983 reported at Tomlinson Run State Park by N. Laitsch (E-Bird 2013). Extra care should be taken during late spring to search for this species at Castleman's Run Lake, and Bear Rocks Lakes, as it could still periodically be present.

Sora *(Porzana carolina)* – Casual Migrant historically. No records post 1948 in the Northern Panhandle, any observation of this bird should be reported as soon possible.

Common Gallinule *(Gallinula galeata)*- Accidental. Two were observed at Bear Rock Lakes on May 7. 1955 (Buckelew, 1976).

American Coot (*Fulica americana*) – Rare migrant. Found

along the Ohio River and lakes in the Panhandle. Ohio River near Wheeling and Wellsburg have the most reports.

Recurvirostridae
American Avocet (*Recurvirostra americana*) – Accidental. Two records. The record from July 6. 1977 was of an injured bird was caught and taken to the Zoo at Ogelbay Park in Wheeling, the individual died a few days later. The Second record in the Panhandle is from 2002 at Shenk Lake and was reported by S.Shalway on September 25th (E-Bird 2013).

Charadriidae
Black-bellied Plover *(Pluvialis squatarola)* – Casual migrant. There are several records from May in Hancock County (Buckelew, 1976).

Semipalmated Plover (*Charadrius semipalmatus*) – Casual migrant. Only a few records in the Panhandle exist may be more numerous and just lacking reports. Found along the Ohio River in existing reports.

Killdeer *(Charadrius vociferus)* – Common migrant and summer resident. Possibly permanent resident as more recent reports from the Panhandle and surrounding area indicate this species is now seen almost year-round. Nests in gravel area such as drive ways. Can be found sometimes on flat rooftops such as the Bethany College Athletic fields where nests are fairly common every summer.

Scolopacidae
Spotted Sandpiper *(Actitis macularius)* – Uncommon summer resident. Nests along streams and some rural areas. I saw a possible breeding pair several times in Bethany

along the Buffalo River behind the College stables in May, 2013.

Solitary Sandpiper *(Tringa solitaria)* – Casual mirgant and summer visitor. No reports are available through E-bird but one from 1948. Dr. Buckelew listed this species as an Uncommon migrant and summer visitor. This could still be true, however with no new data this species could also now have fallen into the Accidental category post 1940's.

Greater Yellowlegs *(Tringa melanoleuca)* – Casual migrant possibly now Accidental. Four records of this bird exist through E-bird with the last being from May, 1968 in Weirton. All other records are from the now nonexistent Beech Bottom Swamp. Extra care should be taken to look for this and other species at wetlands and lakes throughout the Northern Panhandle.

Willet *(Tringa semipalmata)* – Accidental. Charles Conrad reported one at Beech Bottom Swamp in the spring of 1935 (Buckelew, 1976).

Lesser Yellowlegs *(Tringa flavipes)* –Casual migrant. Was somewhat more common than the Greater Yellowlegs and was found along the shores of the Ohio River, and large streams. Especially partial to shallow ponds and mudflats. However, like the Greater Yellowlegs, no records in almost 40 years exists in the Panhandle.

Upland Sandpiper *(Bartramia longicauda)* – Rare migrant and summer resident which will sometimes be heard at night in rural areas. Has been known to nest in Hancock County. However, according to the West Virginia Breeding Bird

Atlas II, there are no records in the Northern Panhandle for this bird breeding nor where there any probable/possible breeding pairs observed.

Whimbrel *(Numenius phaeopus)* – Accidental. One report from July 24, 1933 by George Sutton (G. Sutton, 1933).

Hudsonian Godwit *(Limosa haemastica)* – Accidental. One record by Charles Conrad at Benwood in Marshall County in 1968 (Conrad, 1968).

Ruddy Turnstone (*Arenaria interpres*) – Casual summer migrants. Reports though sporadic are available throughout the Panhandle. The latest record was reported from Ohio County in 1992 where an individual was observed by Dr. Buckelew (E-bird, 2013).

Stilt Sandpiper *(Calidris himantopus)* – Accidental. One was captured and photographed at Wheeling Island on August 10, 1952 (Buckelew, 1976).

Dunlin (*Calidris alpina*) – Accidental/Casual migrant.

Baird's Sandpiper *(Calidris bairdii)* – Accidental. No records past 1936 in the Panhandle exist. Only three records from the 1930's exist, two of which are from Beech Bottom Swamp (E-bird, 2013).

Least Sandpiper *(Calidris minutilla)* – Casual migrant. Prefers mudflats (Buckelew, 1976).

White-rumped Sandpiper *(Calidris fuscicollis)* – Accidental.

Two records exist in the Pandandle. Both records are from Beech Bottom Swamp (E-bird, 2013).

Pectoral Sandpiper *(Calidris melanotos)* – Casual migrant. Species prefers open marshes and mudflats.

Semipalmated Sandpiper *(Calidris pusilla)* – Accidental/Casual migrant. The West Virginia record was established at Beech Bottom Swamp on Sept. 4, 1933 (Buckelew, 1979.)

Short-billed Dowitcher *(Limnodromus griseus)* – Accidental/Casual migrant. Numerous historical reports from Beech Bottom Swamp in the 1930's, however I could find no new reports that exist for Panhandle.

Wilson's Snipe *(Gallinago gallinago)* – Uncommon migrant along the Ohio River and large streams.

American Woodcock *(Scolopax minor)* – Fairly common migrant and early summer resident. Nests have been confirmed in Hancock county (WV Breeding Bird Atlas II, G. Eddy 2013).

Red Phalarope *(Phalaropus fulicarius)* – Accidental. Dead specimen was found by George Breiding on Nov. 4, 1961 at Ogelbay Park (Buckelew, 1976).

Laridae
Bonaparte's Gull (*Chroicocephalus philadelphia*) – Casual/Rare migrant along the Ohio River.

Ring-billed Gull *(Larus delawarenisis)* – Casual migrant along the Ohio River.

Herring Gull *(Larus argentatus)* – Accidental/ Casual migrant along the Ohio River. Often seen with Ring-billed Gulls when bad weather is present.

Great black-backed Gull *(Larus marinus)* – Accidental. Charles Conrad observed one flying over the Ohio River on April 9, 1957. The Second record is from Tom Shields at Oglebay Park on May 11, 1955.

Sooty Tern *(Onychoprion fuscatus)* – Accidental. One record from Aug, 1926 in Wheeling along the Ohio River (E-bird, 2013).

Caspian Tern *(Hydroprogne caspia)* – Casual migrant along the Ohio River.

Black Tern *(Childonias niger)* – Casual migrant along the Ohio River and lakes.

Common Tern *(Sterna hirundo)* – Rare migrant along the Ohio River.

Columbidae
Rock Pigeon *(Columba livia)* – Very common permanent resident. Nests on buildings, under bridges, and in barns. Most often is considered a pest.

Mourning Dove *(Zenaida macroura)* – Common permanent resident. Sometimes seen in large flocks in fields and is a common feeder bird.

Cuculidae

Yellow-billed Cuckoo *(Coccyzus americanus)* – Uncommon summer resident.

Black-billed Cuckoo *(Coccyzus erythropthalmus)* – Uncommon summer resident, more common in both the Northern and Southern parts of the Panhandle.

Tytonidae
Barn Owl (*Tyto alba*) – Rare summer resident. Less common in winter. Will use barns, hollow trees, and empty buildings for nesting sites. I found a possible nesting pair in Bethany, WV on March, 29, 2013 that was later confirmed in the area by state biologists.

Strigidae
Eastern Screech-Owl *(Megascops asio)* – Common permanent resident. Found in woods, orchards, towns all throughout the Panhandle.

Great Horned Owl *(Bubo virginianus)* - Uncommon permanent resident. Found in more remote wooded areas, especially when streams are present. Bethany Park has had Great Horned Owls from time to time.

Snowy Owl *(Bubo scandiacus)* - Accidental/Casual winter visitor. Most often found on fence posts, or trees in open countries in years it is present in the Pandhandle.

Barred Owl *(Strix varia)* - Uncommon permanent resident of remote, upland woods.

Long-eared Owl *(Asio otus)* - Casual migrant and winter visitor

Short-eared Owl *(Asio flammeus)* - Casual migrant and

winter visitor

Northern Saw-whet Owl *(Aegolius acadicus)* - Accidental. I have called two individuals in very near me once in the Bethany area by the Gresham Inn on March 29, 2013. Both individuals were seen approx. a week later, and haven't been seen since. Most likely this was a pair passing through the Panhandle.

Caprimulgidae
Common Nighthawk *(Chodeiles minor)* – Fairly common fall migrant and uncommon summer resident of both large towns and cities, will nest on gravel roof tops.

Eastern Whip-poor-will *(Antrostomus vociferus)* - Casual migrant and rare summer resident.

Apodidae
Chimney Swift *(Chaetura pelagica)* - Common summer resident. Nests in chimneys and large hollow tree. Bethany College is home to a large amount around and in the chimneys of Old Main.

Trochilidae
Ruby-throated Hummingbird *(Archilochus colubris)* – Fairly common summer resident. Nests in woods, and sometimes bushes near homes. Feeds on flower nectar and when fed by flowers or feeders can become a regular visitor.

Rufous Hummingbird (*Selasphorus rufus*) – Casual summer migrant. Formerly not found in the Panhandle, this species has become a casual migrant in both the Panhandle and surrounding area.

Alcedinidae
Belted Kingfisher *(Ceryle alcyon)* – Fairly common summer resident. Less common in winter. Will nest in burrows made

in banks near streams, found in winter until the streams freeze over. Buffalo Creek near Bethany is an excellent place to find Belted Kingfisher, especially near the soccer practice fields.

Picidae

Red-headed Woodpecker *(Melanerpes erythrocephalus)* – Rare permanent resident. While this bird is rare it is making significant leaps in population throughout the state according to the new Breeding Bird Atlas data. Recent reports have spanned most of the Panhandle however, Marshall county in the South according to reports is the best place to look.

Red-bellied Woodpecker (*Melanerpes carolinus*) – Fairly common permanent resident. Found in woods, towns, and a common feeder if a suet feeder is present.

Yellow-bellied Sapsucker (*Sphyrapicus varius*) – Uncommon migrants and winter visitor. Leaves rows of holes in tree trunks from which is drinks the sap, occasionally will even visit suet feeders. I observed this bird in Bethany during the winter of 2013.

Downy Woodpecker (*Picoides pubescens*) – Common permanent resident found in wooded areas. Common suet feeder.

Hairy Woodpecker (*Picoides villosus*) – Fairly common permanent resident.

Black-backed Woodpecker *(Picoides articus)* – Accidental. One was seen by E.R. Chandler near Chest, WV in 1962.

Northern Flicker (*Colaptes auratus*) – Common summer resident possibly permanent resident. Reports since 1979's publication indicate that not all Flickers may be migrating as most months of the year there are reports that exist from most of the Panhandle.

Pileated Woodpecker (*Dryocopus pileatus*) – Fairly common permanent resident of woods, parks and towns in the Panhandle. According to the first publication of this guide, George Sutton reported that this bird was absent from the Panhandle prior to 1933 (G. Sutton, 1933). In 1938 this bird is also absent from the list of Ogelbay Park birds. However, as with much of the Panhandle, this bird can now be found here. One excellent spot for this bird is the woods of Bethany College. Especially in the old growth portion deeper into the trails.

Falconidae
American Kestrel (*Falco sparverius*) – Uncommon permanent resident. Found hovering over fields or perched on power lines along the road.

Peregrine Falcon (*Falco peregrinus*) - Casual migrant. Was once more common than it is now.

Tyrannidae
Olive-sided Flycatcher (*Contopus cooperi*) – Casual migrant. I have only seen one of these in 4 years in the Panhandle.

Eastern Wood-Pewee *(Contopus virens)* – Common summer resident found in woodlands.

Yellow-bellied Flycatcher *(Empidonax flaviventris)* – Uncommon migrant. Fall is the more common season for this bird, is often seen in Willow trees along creeks, and streams.

Acadian Flycatcher (*Empidonax virescens*) – Common summer resident found on steep wooded hillsides. I have had luck some years in Parkinson Forrest finding this species.

Alder Flycatcher (*Empidonax alnorum*) – See Willow Flycatcher.

Willow Flycatcher (*Empidonax traillii*) - Rare summer resident easily confused with the Willow Flycatcher during migration. Bear Rocks Lakes is a possible location to observe this bird (Buckelew, 1976).

Least Flycatcher (*Empidonax minimus*) -Uncommon migrant. I have only seen this bird twice in the Panhandle, both times in Parkinson Forest.

Eastern Phoebe (*Sayornis phoebe*) – Common summer resident. Will build nests under bridges, waterfalls, and cliff over hangs.

Vermillion Flycatcher *(Pyrocephalaus rubinus)* – Accidental. One was reported by Mrs. Richard Jennings at her home around May 12, 1956 (Buckelew, 1976).

Great Crested Flycatcher (*Myiarchus crinitus*) – Fairly common summer visitor. Good locations to find this bird are Bethany College Forest (Parkinson Forest), Tomlinson Run State Park, and Brooke Hills Park. Listen for its call which is a "creep".

Eastern Kingbird (*Tyrannus verticalis*) – An uncommon summer resident. A good spot to find this bird is the Bethany College stables, follow the fence line behind Campbell Manor until the two sections meet and look for them on the fence, this also a good location for Red-winged Black birds, and Yellow Warblers, and some waterfowl due to the Buffalo Creeks close proximity.

Laniidae
Loggerhead Shrike *(Lanius ludovicianus)* – Casual Migrant.

Northern Shrike (*Lanius excubitor*) – Casual migrant.

Vireonidae
White-eyed Vireo *(Vireo griseus)* – Uncommon summer resident. Found in brushy ravines, old fields, and along road sides. This species along with the Yellow-throated, Warbling, and Red-eyed Vireo all breed in the Northern Panhandle (WV Breeding Bird Atlas II, G. Eddy 2013).

Yellow-throated Vireo *(Vireo flavifrons)* – Fairly common summer resident. Found in open woods. Has been known to breed on Middle Wheeling Creek (Buckelew, 1976).

Blue-headed Vireo *(Vireo solitarius)* – Uncommon summer resident. I have only seen this species once in the Panhandle.

Warbling Vireo *(Vireo gilvus)* – Fairly common summer resident. Can be found along roads, streams and in towns throughout the Panhandle.

Philadelphia Vireo *(Vireo philadelphicus)* – Casual/Rare migrant.

Red-eyed Vireo *(Vireo olivaceus)* – Common summer resident. Found in woodlands.

Corvidae
Blue Jay (Cyanocitta cristata) – Fairly common summer resident and winter visitor.

American Crow (Corvus brachyrhynchos) – Common permanent resident.

Common Raven (Corvus corax) – Rare permanent resident. Few records exist however, possible breeding pairs were found in Marshall County recently for the West Virginia Breeding Bird Atlas II (WV Breeding Bird Atlas II, G. Eddy 2013).

Alaudidae
Horned Lark *(Eremophila alpestris)* – Fairly common permanent resident of large grassy hilltops (Buckelew, 1976). According to Dr. Buckelew the best spots to observe the Horned Lark are Waterford Downs, Brooke Hills Park, and the Ohio County Airport.

Hirundinidae
Purple Martin *(Progne subis)* – Uncommon summer resident. Martin houses are located on Gertie's Point Road in West Liberty.

Tree Swallow *(Tachycineta bicolor)* – Rare migrant. Often observed over lakes. I have seen this species on just two occasions in Bethany.

Northern Rough-winged Swallow *(Stelgidopteryx serripennis)* – Common summer resident. Nests in the banks of the Buffalo Creek. Will sometimes use old Belted Kingfisher burrows for its nest.

Bank Swallow *(Riparia riparia)* – Rare migrant.

Cliff Swallow *(Petrochelidon pyrrhonota)* – Casual migrant.

Barn Swallow *(Hirundo rustica)* – Common summer resident. Nest can be found in barns, under eaves or porches.

Paridae
Carolina Chickadee *(Poecile carolinensis)* – Common permanent resident. Virtually indistinguishable from the Black-capped Chickadee in the field. Call is a whistled four syllable "Fee-bee, fee-bay".

Black-capped Chickadee *(Poecile atricapillus)* – Uncommon winter visitor. Very possible that this is found all year due to the similarity in features of the Carolina Chickadee. Call is "Chick-a-dee-dee-dee" or a clear whistle of "fee-bee-ee or fee-bee"

Tufted Titmouse *(Baeolophus bicolor)* – Common permanent resident. Will breed in woodlands until winter arrives when it will then move into towns and feeds at bird feeders. Commonly seen with Chickadees and Nuthatches at feeders.

Sittidae
Red-breasted Nuthatch *(Sitta canadensis)* – Uncommon migrant and winter visitor.

White-breasted Nuthatch (*Sitta carolinensis*) – Common permanent resident.

Certhiidae
Brown Creeper *(Certhia americana)* – Fairly common migrant. Rare winter visitor and rare summer resident.

Troglodytidae
House Wren (*Troglodytes aedon*) – Common summer resident.

Winter Wren (*Troglodytes hiemalis*) – Uncommon migrant and winter visitor.

Sedge Wren (*Cistothorus platensis*) – Accidental/ Possible Casual migrant. One specimen was collected by George Sutton at Bethany on September of 1936 (Sutton, 1937).

Marsh Wren (*Cistothorus palustris*) – Accidental/ Possible Casual migrant. Karl Haller once collected several specimens from Beech Bottom Swamp (Buckelew, 1976).

Carolina Wren (*Thryothorus ludovicianus*) – Common permanent resident. Can be found in both woods and brushy fields.

Bewick's Wren *(Thryomanes bewickii)* – Accidental/ Casual migrant. Prior to the 1979 edition this was a uncommon summer resident.

Polioptilidae
Blue-gray Gnatcatcher *(Polioptila caerulea)* – Uncommon summer resident. Found in woodlands and parks. As with several others, I have found this species around Campbell Manor in Bethany.

Regulidae

Golden-crowned Kinglet (*Regulus satrapa*) – Fairly common winter visitor. Look for large stands of trees which this bird prefers to improve observation chances.

Ruby-crowned Kinglet (*Regulus calendula*) – Fairly common migrant and uncommon winter visitor.

Turdidae

Eastern Bluebird (*Sialia sialis*) – Fairly common permanent resident. Competes with American Robins over nesting spots including nest boxes.

Veery (*Catharus fuscescens*) – Casual migrant. I have only found this bird one time in all my field walks in Bethany College Forest.

Gray-cheeked Thrush *(Catharus minima)* – Casual/Rare migrant.

Bicknell's Thrush (*Catharus bicknelli*) – Accidental. G. Breiding reported one hitting the window which resulted in its death at the Nature Center of Oglebay Park on September 19, 1961.

Swainson's Thrush (*Catharus ustulatus*) – Rare formerly Uncommon migrant. I have only found one report that was from G.Breiding at Oglebay Park on April 25, 1960. However, the surrounding area of the Panhandle has dozens of reports, which leads me to believe this bird is present but not observed/reported.

Hermit Thrush (*Catharus guttatus*) – Uncommon migrant and casual winter visitor.

Wood Thrush (*Hylocichla mustelina*) – Common summer

resident.

American Robin (*Turdus migratorius*) – Very common summer resident. Uncommon winter visitor. During migration can be extremely abundant when present.

Mimidae
Gray Catbird *(Dumetella carolinensis)* – Common summer resident and casual winter visitor. Can be found in shrubs, towns, yards of farms, and rows of hedges.

Brown Thrasher *(Toxostoma rufum)* – Fairly common summer resident of brushy edges.

Northern Mockingbird *(Mimus polyglottos)* – Fairly common summer visitor, rare winter visitor. Some individuals may stay through the winter as I have seen and photographed several in the Bethany area that keep a territory year-round.

Sturnidae
European Starling *(Sturnus vulgaris)* – Abundant permanent resident.

Motacillidae
American Pipit *(Anthus rubescens)* – Casual migrant. Usually seen in spring. I have seen Pipits in the Practice field near Buffalo Creek in Bethany. Not far from my reported location in April 1934, George Sutton reported a flock of over 100 (Sutton, 1935).

Bombycillidae
Cedar Waxwing *(Bombycilla cedrorum)* – Fairly common permanent resident.

Calcariidae
Lapland Longspur *(Calcarius lapponicus)* – Accidental. R.

Matesic reported seeing this species along with Snow Buntings on February 2. 1979, on Fork Ridge in Marshall county along route 17.

Snow Bunting *(Plectrophenax nivalis)*- Casual winter migrant. When present flocks that contain dozens of birds can sometimes be seen.

Parulidae
Ovenbird *(Seiurus aurocapilla)* – Fairly common summer resident in woodlands.

Worm-eating Warbler *(Helmitheros vermivorum)* – Uncommon summer resident found in heavy woods throughout the Panhandle.

Louisiana Waterthrush *(Parkesia motacilla)* – Fairly common spring and summer resident. Can be found near small, swift streams in woodlands.

Northern Waterthrush (*Parkesia noveboracensis*) – Rare migrant.

Golden-winged Warbler *(Vermivora chrysoptera)* – Rare summer resident, rare winter and spring visitor. Records have shown this species to show up in late winter, and early spring. I have seen this bird as early as April on one occasion.

Blue-winged Warbler (*Vermivora cyanoptera*) – Common summer resident. Found in brushy fields. Now more common than in 1979's publication which was more common than its previous publication in the 1930's. Dr. Buckelew notes that Brewster's Warbler a hybrid between the Blue-winged and Golden-winged Warbler, is possible in the Panhandle. Not only has this hybrid been observed, one even had a territory near West Liberty from 1974 until 1975. No nest was found however, but care should still be taken to locate both this hybrid and nests.

Black-and-white Warbler *(Mniotilta varia)* – Uncommon summer resident of woodlands. I have observed this just once in the Panhandle.

Prothonotary Warbler *(Protonotaria citrea)* – Casual migrant.

Tennessee Warbler *(Oreothlypis peregrina)* – Common migrant.

Orange-crowned Warbler *(Oreothlypis celata)* – Casual migrant.

Nashville Warbler *(Oreothlypis ruficapilla)* – Fairly common migrant.

Connecticut Warbler *(Oporornis agilis)* – Accidental/ Casual migrant.

Mourning Warbler *(Geothlypis philadelphia)* – Casual/Rare migrant

Kentucky Warbler (*Geothlypis formosa*) – Fairly common summer resident of woodlands.

Common Yellowthroat *(Geothlypis trichas)* – Uncommon summer resident found in wet fields.

Hooded Warbler *(Setophaga citrina)* – Fairly common summer resident of woods. Can be found at both Ogelbay and Tomlinson Run Park. I have heard the song near Bethany. Listed as a probable breeding species (WV Breeding Bird Atlas II, G. Eddy 2013).

American Redstart *(Setophaga ruticilla)* – Fairly common summer resident of Woodlands. Oglebay, Brooke Hills Park, Tomlinson Run Rarks are very good locations to find the bird. Formerly rare now uncommon in the Bethany area.

Cape May Warbler *(Setophaga tigrina)* – Fairly common migrant and casual winter visitor.

Cerulean Warbler *(Setophaga cerulea)* – Uncommon summer resident of woodlands. Rare in some parts of the Panhandle. Cerulean Warbler populations have been declining due to Winter habitats being destroyed. Bethany College Forest which used to have a population of them, has seen a very sharp decline to the point this species may be extirpated in Parkinson Woods. However, this bird has been reported as probable breeding species (WV Breeding Bird Atlas II, G. Eddy 2013).

Northern Parula *(Setophaga americana)* – Rare migrant. Most E-bird reports are from Marshall county in the Southern part of the Panhandle.

Magnolia Warbler *(Setophaga magnolia)* – Fairly common migrant.

Bay-breasted Warbler *(Setophaga castanea)* – Rare migrant.

Blackburnian Warbler *(Setophaga fusca)* – Uncommon migrant.

Yellow Warbler *(Setophaga petechia)* – Common summer resident. Can be found in towns, parks and farm yards.

Chesnut-sided Warbler (*Setophaga pensylvanica*) – Fairly common migrant.

Blackpoll Warbler *(Setophaga striata)* – Casual migrant.

Black-throated Blue Warbler *(Setophaga caerulescens)* – Uncommon migrant.

Palm Warbler *(Setophaga palmarum)* – Rare migrant

Pine Warbler *(Setophaga pinus)* – Casual migrant

Yellow-rumped Warbler (*Setophaga coronata*) – Common migrant and uncommon winter visitor. The subspecies found in the Northern Panhandle is the Myrtle Warbler.

Yellow-throated Warbler (*Setophaga dominica*) – Uncommon migrant.

Prairie Warbler (*Setophaga discolor*) – Rare migrant. Has been a recorded species to breed in Hancock County (WV Breeding Bird Atlas II, G. Eddy 2013).

Black-throated Green Warbler (*Setophaga virens*) – Casual/Rare migrant

Canada Warbler *(Cardellina canadensis)* –
Accidental/Casual migrant.

Wilson's Warbler *(Cardellina pusilla)* – Casual migrant.

Yellow-breasted Chat *(Icteria virens)* – Rare summer
resident of old fields.

Emberizidae
Eastern Towhee (*Pipilo erythrophthalmus*) - Common
summer resident of woods and edges. Casual winter visitor.

Bachman's Sparrow *(Peucaea aestivalis)* – Accidental. Karl
Haller collected a singing male near West Liberty in July of
1934 (Buckelew, 1976).

American Tree Sparrow *(Spizella arborea)* – Common
winter visitor.

Chipping Sparrow (*Spizella passerina*) – Common summer
resident of parks and towns.

Field Sparrow (*Spizella pusilla*) – Common summer
resident and casual in winter. Found in brushy fields
throughout the Panhandle.

Vesper Sparrow (*Pooecetes gramineus*) – Casual summer
resident of fields and meadows.

Lark Sparrow *(Chondestes grammacus)* – Casual formally
rare summer resident. This species has all but disappeared
from reports since the 1970's.

Savannah Sparrow *(Passerculus sandwichensis)* – Uncommon summer resident. Found in upland fields and can be observed in fields near north of West Liberty and Bethany College.

Grasshopper Sparrow *(Ammodramus savannarum)* – Uncommon formerly common summer resident. Reports indicate that this species has declined in the Panhandle. This species still breeds in the Northern Panhandle despite population declines (WV Breeding Bird Atlas II, G. Eddy 2013).

Henslow's Sparrow *(Ammodramus henslowii)* – Casual/Rare summer resident. Casual in frequency, and rare in years present in the Panhandle.

Le Conte's Sparrow (*Ammodramus leconteii*) – Accidental. Sutton and Haller's two Beech Bottom Swamp records are still the only records in the state of West Virginia according to E-bird.

Nelson's Sparrow (*Ammodramus nelsoni*) – Accidental. Only two records from Beech Bottom Swamp in 1948 by Karl Haller exist.

Fox Sparrow *(Passerella iliaca)* – Casual migrant and winter visitor

Song Sparrow (*Melospiza melodia*) – Common permanent. Found in towns, brushy edges, towns and parks.

Lincoln's Sparrow (*Melospiza lincolnii*) – Casual migrant. In years it is present in the Panhandle large flocks of 100 or more birds may be observed.

Swamp Sparrow (*Melospiza georgiana*) – Casual migrant.

White-throated Sparrow (*Zonotrichia albicollis*) – Uncommon migrant and fairly common winter visitor.

Harris's Sparrow (*Zonotrichia querula*) – Accidental. Dr. Buckelew observed one closely in a power line cut on February 26, 1971 while in Bethany.

White-crowned Sparrow (*Zonotrichia leucophrys*) – Uncommon migrant. More common in the spring.

Dark-eyed Junco (*Junco hyemalis*) – Common winter visitor.

Cardinalidae
Summer Tanager (*Piranga rubra*) – Casual summer visitor. As of the 1979 edition of this guide, only two records exist. This number has now quadrupled through E-Bird.

Scarlet Tanager (*Piranga olivacea*) – Fairly common summer resident of mature woodlands.

Northern Cardinal (*Cardinalis cardinalis*) – Common permanent resident of woodlands and edges.

Rose-breasted Grosbeak (*Pheucticus ludovicianus*) – Fairly common migrant. Rare summer resident. Has nested in

Tomlinson Run State Park (Buckelew, 1976).

Black-headed Grosbeak (*Pheucticus melanocephalus*) –
Accidental. Charles Conrad reports seeing one on December
9, 1979 in Triadelphia in Ohio County.

Blue Grosbeak (*Passerina caerulea*) – Casual migrant

Indigo Bunting *(Passerina cyanea)* – Common summer
resident of roadsides, parks, the edges of woodlands. One of
the most common roadside birds.

Dickcissel *(Spiza americana)* – Casual migrant.

Icteridea
Bobolink (*Dolichonyx oryzivorus*) – Uncommon summer
resident in Brooke and Hancock counties. Migrant to the
south. West Virginia's first recorded nest was found in
Hancock County by Robert Murray.

Red-winged Blackbird (*Agelaius phoeniceus*) – Common
summer resident. Found in wet fields, cultivated alfalfa, and
pond margins.

Eastern Meadowlark (*Sturnella magna*) – Common summer
resident of hayfields and pastures. Casual winter visitor.
Look for this bird on stumps, power lines, fences, and crests
of hills in appropriate habitat.

Yellow-headed Blackbird (*Xanthocephalus xanthocephalus*)
– Accidental. Sutton reported a male on June 16, 1914

flying over Bethany (Buckelew, 1976).

Rusty Blackbird (*Euphagus carolinus*) – Casual migrant.

Brewer's Blackbird (*Euphagus cyanocephalus*) –
Accidental. Dr. Buckelew reports seeing this bird in
Bethany in December of 1980.

Common Grackle (*Quiscalus quiscula*) – Very common
summer resident. Nests in large pine trees. Large flocks of
this bird can be found during migration in pine trees,
uncommon in winter.

Brown-headed Cowbird (*Molothrus ater*) – Common
summer resident. Very common migrant. Casual winter
visitor.

Orchard Oriole (*Icterus spurius*) – Uncommon summer
resident. Found in towns, orchards and parks. Range now
extends throughout the entire Panhandle.

Baltimore Oriole (*Icterus galbula*) – Common summer
resident of towns, and parks.

Fringillidae
Pine Grosbeak (*Pinicola enucleator*) – Accidental winter
visitor.

House Finch (*Haemorhous mexicanus*) – Common
permanent resident. Nests have now been confirmed in
almost every county in the Panhandle.

Purple Finch (*Haemorhous purpureus*) – Casual winter visitor. Can appear in large flocks of 30 or more.

Red Crossbill *(Loxia curvirostra)* – Casual winter visitor. Often will be seen in flocks of 25 or more birds (Buckelew, 1976).

White-winged Crossbill (*Loxia leucoptera*) – Casual winter visitor.

Common Redpoll (*Acanthis flammea*) – Casual winter visitor. Along the Ohio River flocks of over 100 birds can be seen.

Pine Siskin (*Spinus pinus*) – Casual migrant and winter visitor.

American Goldfinch (*Spinus tristis*) – Common permanent resident. Common at feeders. Nests can be found during late summer in brushy fields and edges.

Evening Grosbeak *(Coccothraustes vespertinus)* – Casual winter visitor. When present will visit sunflower feeders.

Passeridae
House Sparrow *(Passer domesticus)* – Very common to Abundant permanent resident. Depending on availability of both food and nesting sites, this bird can be found at nearly every home. Considered a pest species.

LITERATURE CITED
"AOU Checklist of North and Middle American Birds."

American Ornithologists' Union, 13 Sept.   2013. Web. 23 Sept. 2013.

American Ornithologists' Union, 2013. 54th Supplement (Auk 2013, Vol. 130:558-571), Baltimore, Md.

Beatty, W.H. 1975. Brewster's Warbler in Ohio County, West Virginia. **Redstart**, 42:121

Breiding, G.H. 1953. Some recent miscellaneous notes from Ohio County, West Virginia. **Redstart**, 20:29-30.

Breiding, G.H. 1955a. A nesting record for the Long-eared Owl in West Virginia. **Redstart**, 22:36.

Breiding, G.H. 1955b. An Ohio County record for the White-winged Cross bill. **Redstart**, 22:27.

Breiding, G.H. 1959. Blue Jay nesting in Ohio County. **Redstart**, 26:55

Breiding, G.H. 1962a. Red Phalarope in West Virginia. **Wilson Bull**, 74:288.

Breiding, G.H. 1962b. Summer Tanager in Ohio County. **Redstart**, 29:17

Brooks, A.B. 1938. The Trail Guide. Oglebay Institute, Institute Edue. Bull. 2, Wheeling, W.Va.

Brooks, M.G. 1944. A Check-list of West Virginia Birds. West Virginia Univ. Agri. Exp. Bull. 316.

Buckelew, Jr., Albert R. 1971. Breeding Bird Census: mature northern hardwoods. **Amer. Birds,** 25:972.

Buckelew, Jr., Albert R. *Birds of the West Virginia Northern Panhandle*. St Albans, WV: Harlass Printing, 1976. Print.

Buckelew, Jr., Albert R. *Birds West Virginia Field Checklist*. St Albans, WV: WVDNR, 2004. Print.

Conrad, C.L. 1960. Raven reported at Wheeling. **Redstart**, 27:42.

Conrad, C.L. 1968. Hudsonian Godwit is West Virginia record. **Redstart**, 35:59.

Eddy, Greg E. *Birding Guide to West Virginia*. 2nd ed. Wheeling, W. Va.: Club, 2009. Print.

"E-Bird." *E-Bird Database*. Cornell Lab of Ornithology, 1 Jan. 2012. Web. 12 Dec. 2012.

Ellyson, W.J., Kunkle, M., Ruffner, J.D., and Webb, J. 1974. Soil Survey of Brooke, Hancock, and Ohio Counties, West Virginia. U.S. Dept. of Agricultural.

"Geology.com: News and Information for Geology & Earth Science." *West Virginia County Map with   County Seat Cities*. Web. 23 Oct. 2014.

Gorman, M. 1954. Summer Tanager in Marshall County. **Redstart**, 21:45.

Haller, K.W. 1935. Notes from Bethany, West Virginia. **Cardinal**, 4:49-50.

Haller, K.W. 1961. Final notes on Beech Bottom Swamp. **Redstart**, 28:96-101.

Hurley, G.F. 1960. Field Notes, **Redstart**, 27:65-76.

Hurley, G.F. 1962. Field Notes, **Redstart**, 24:83.

Jennings, W. 1954. Mockingbird nests in Northern Panhandle. **Redstart**, 21:64-65.

Katholi, C. 1971. The Gathering Cage. **Redstart**, 38:22.

Laitsch, J. and Chandler, E.R. 1975, Black-backed three-toed woodpecker seen in Ohio. **Redstart**, 42:128

Laitsch, N. 1964. Field Notes. **Redstart**, 31:91-95.

Laitsch, N. 1966. Field Notes. **Redstart**, 33:58-62.

Laitsch, N. 1969a. Field Notes. **Redstart**, 36:91-93.

Laitsch, N. 1969b. Field Notes. **Redstart**, 36:110-116.

Laitsch, N. 1971. Field Notes. **Redstart**, 38:102-105.

Laitsch, N. 1974. Field Notes. **Redstart**, 41:121-125.

Laitsch, N. 1975. Field Notes. **Redstart**, 42:89-92.

Montagna, W. 1940. Bald Eagle in the West Virginia Panhandle. **Cardinal**, 5:68-69.

Montagna, W. 1940. King Rail in the West Virginia Panhandle. **Cardinal**, 5:70.

"Northeast Regional Climate Center." Web. 23 Oct. 2014.

Phillips, G.F. 1957. An area study, Ohio County, West Virginia. **Redstart**, 24:48-52.

Phillips, G.F. 1973. Winter and breeding bird population studies at Bear Rocks Lakes. **Redstart**, 40:51-53.

Sutton, G.M. 1933, Birds of the West Virginia Panhandle, **Cardinal**,3:101-124

Sutton, G.M. 1934. Hudsonian Curlew in the W. Va. Panhandle. **Cardinal**, 3:101-124.

Sutton, G.M. 1935. Spring notes from Bethany, W. Va. **Cardinal**, 4:50.

Sutton, G.M. 1937. Notes from Brooke County, West Virginia. **Cardinal**, 4:117-118.

Temple, P. and Temple, F. 1974. Effect of habitat changes on water and shore birds. **Redstart**, 41:114-119.

Vossler, B. 1967. Brown Creepers nesting in Ohio County. **Redstart**, 34:74

Weimer, B.R. 1935. Holboell's Grebe at Bethany, W. Va. **Cardinal**, 4:17.

West, R. and Shields, T. 1935. Some bird records for the Norhern W. Va. Panhandle. **Redstart**, 2:23-28.

"WVDNR Website - West Virginia Division of Natural Resources." West Virginia Department of   Natural Resources, 1 Jan. 2012. Web. 23 Dec. 2012.

Yenke, W.H. 1952. Stilt Sandpiper in Ohio County, West

Virginia. **Redstart**, 20:18.

www.ingramcontent.com/pod-product-compliance
Lightning Source LLC
Chambersburg PA
CBHW071828200526
45169CB00018B/1187